Eye on the Environment Worksheets with Answer Key

HOLT, RINEHART AND WINSTON
Harcourt Brace & Company

Austin • New York • Orlando • Atlanta • San Francisco • Boston • Dallas • Toronto • London

ISBN 0-03-051447-9   2345   021   00   99   98   97

Contents

About Eye on the Environment Worksheets   iv

Unit 1: Mapping Life on Earth   1

Unit 2: Volcanoes Put a Chill in the Air   3

Unit 3: What's Mined Is Yours   5

Unit 4: Losing Ground   7

Unit 5: The Mysterious Bone Bed   9

Unit 6: A Harbor Makes a Comeback   11

Unit 7: Keeping Cool With Algae   13

Unit 8: The Great Junkyard in the Sky   15

Answer Key   17

About Eye on Environment Worksheets

Eye on the Environment Worksheets is designed for use with the Eye on the Environment features in *Modern Earth Science*. These worksheets are blackline masters, which can be reproduced and distributed to students. Each worksheet is divided into three parts. The first part contains questions to help check students' reading comprehension. The second and third parts contain questions that require students to make inferences based on the feature or to express their own ideas about the topic covered in the feature. Sample answers are provided at the end of this booklet to help you assess students' responses.

M O D E R N   E A R T H   S C I E N C E

Unit 1: Studying the Earth

Eye on the Environment: Mapping Life on Earth

[Use this worksheet with pages 62–63 of your textbook.]

Identifying the Facts

1. What was special about the animals that were discovered in Southeast Asia?

2. Why were the animals in danger?

3. Name at least two methods that scientists use to identify ecosystems that may be in need of protection.

4. Name some types of areas with a high level of biodiversity.

5. What is a biodiversity map?

Interpreting the Story

6. In your own words, define the term *biodiversity*.

7. Name some types of areas that might have a low level of biodiversity.

____

____

____

8. How do you think satellite photographs are used to create a biodiversity map?

____

____

____

____

____

____

Thinking a Little Harder

9. How is creating a biodiversity map different from creating a topographic map? How is it similar?

____

____

____

____

____

10. Could you create a biodiversity map that is also topographic? Why or why not?

____

____

____

____

____

____

M O D E R N   E A R T H   S C I E N C E

Unit 2: The Dynamic Earth

Eye on the Environment:
Volcanoes Put a Chill in the Air

[Use this worksheet with pages 134–135 of your textbook.]

Identifying the Facts

1. Why do some scientists think there is a link between human activities and the earth's natural greenhouse effect?

____

____

____

____

2. What chemical change happens to sulfur dioxide once it is released into the atmosphere?

____

____

3. What were the effects of Mount Pinatubo's eruption on the temperature of the earth?

____

____

____

4. Why did the eruption of Mount Pinatubo cause the average global temperature to change?

____

____

____

____

Interpreting the Story

5. Why are studies of volcanoes helpful to scientists who wish to understand the greenhouse effect?

____

____

____

____

____

____

M O D E R N   E A R T H   S C I E N C E

6. After Mount Pinatubo erupted, the effects on the earth's temperature were not observed until more than a year had passed. Explain why the changes did not occur immediately.

____

____

____

7. Many scientists observed the effects of Mount Pinatubo's eruption on global climate. Form a hypothesis that these scientists might have been testing with these data.

____

____

____

____

8. Why did the scientists monitor average global surface temperatures instead of temperatures at a specific location?

____

____

____

____

____

9. What are some other natural phenomena that scientists might observe in order to study global climate changes?

____

____

____

____

Thinking a Little Harder

10. How would you determine whether the 1991 Mount Pinatubo eruption affected the average monthly temperatures in your community?

____

____

____

____

____

____

____

4

M O D E R N   E A R T H   S C I E N C E

Unit 3: Composition of the Earth

Eye on the Environment: What's Mined Is Yours

[Use this worksheet with pages 214–215 of your textbook.]

Identifying the Facts

1. If the environmental costs of mining operations are so high, why do the mines exist?

____

____

____

2. What are some of the resources that come from mines?

____

____

____

3. Name at least one negative short-term effect of a mining operation.

____

____

____

____

4. Name at least one negative long-term effect of a mining operation.

____

____

____

5. What are some organisms that are used in reclamation efforts?

____

____

6. Why do mining companies sometimes go out of business?

____

____

____

____

Interpreting the Story

7. Besides mining operations, what other kinds of activities might necessitate reclamation?

____

____

____

____

8. Why are living organisms helpful in reclamation efforts?

____

____

____

____

9. Why does recycling help to lessen the problems associated with mining?

____

____

____

____

____

Thinking a Little Harder

10. In areas where old mines have never been reclaimed, the original mining companies have often long been out of business. In these cases, who do you think should have to pay the costs of reclaiming the mines?

____

____

____

____

____

____

M O D E R N   E A R T H   S C I E N C E

Unit 4: Reshaping the Crust
Eye on the Environment: Losing Ground

[Use this worksheet with pages 318–319 of your textbook.]

Identifying the Facts

1. Why are concerns about food shortages becoming more serious with each passing day?

2. On a small scale, how does the erosion of topsoil occur?

3. Why is soil in a forest less likely to be eroded than soil in an open, agricultural field?

4. Where does soil go once it has been dislodged from its original location?

5. Name four agricultural techniques that can reduce erosion in fields.

Interpreting the Story

6. Which is less susceptible to erosion, a natural grassland or a suburban front yard? Explain your answer.

7. How could a suburban front yard be changed to reduce the risk of erosion?

____

____

____

____

____

8. Would erosion be greater on a level surface or a sloped surface (such as a field on the side of a hill)? Explain your answer.

____

____

____

____

____

____

____

Thinking a Little Harder

9. What are some factors that might keep a farmer from using sustainable farming methods?

____

____

____

____

____

____

____

____

M O D E R N   E A R T H   S C I E N C E

Unit 5: The History of the Earth

Eye on the Environment: The Mysterious Bone Bed

[Use this worksheet with pages 384–385 of your textbook.]

Identifying the Facts

1. Why is the bone bed referred to as a fossil treasure chest?

2. Beside mammals, what other types of organisms are represented by the Kipsaramon fossils?

3. If a scientist discovered a fossil skeleton that was articulated, how would it be different from the fossils found at Kipsaramon?

4. Why are the fossils shipped to Nairobi?

5. What is distinctive about fossils that have been deposited by a predator?

Interpreting the Story

6. Why is it helpful to find a large population of a single species in a bone bed?

M O D E R N   E A R T H   S C I E N C E

7. From the list of organisms that were present at the site, what do you think the environment might have looked like?

____

____

____

____

8. Do you think scientists would be likely to find dinosaur fossils at this site? Explain your answer.

____

____

____

____

Thinking a Little Harder

9. Propose a hypothesis that explains how the Kipsaramon bone bed may have formed. How would you test your hypothesis?

____

____

____

____

____

____

____

____

M O D E R N   E A R T H   S C I E N C E

Unit 6: Oceans

Eye on the Environment:
A Harbor Makes a Comeback

[Use this worksheet with pages 450–451 of your textbook.]

Identifying the Facts

1. For how long was sewage dumped directly into Boston Harbor?

____

____

____

2. What were the effects on harbor organisms?

____

____

____

3. In the 1970's, what facilities were in place to treat sewage from Boston? What facilities were added over the next two decades?

____

____

____

____

____

4. What positive changes in Boston Harbor might a diver observe?

____

____

____

5. What is a watershed?

____

____

____

____

MODERN EARTH SCIENCE

Interpreting the Story

6. How did sewage cause outbreaks of typhoid and cholera among Boston residents?

____

____

____

7. Why do you think the input of scientists was necessary when citizens and public officials wished to correct the problems in Boston Harbor?

____

____

____

____

8. When a federal judge ordered the state of Massachusetts to comply with the Clean Water Act, why do you think the state was given so much time? In other words, why didn't the judge order them to comply immediately?

____

____

____

____

____

Thinking a Little Harder

9. Why would the health of Boston Harbor depend on the health of the surrounding watersheds?

____

____

____

____

____

____

M O D E R N   E A R T H   S C I E N C E

Unit 7: Atmospheric Forces

Eye on the Environment: Keeping Cool With Algae

[Use this worksheet with pages 542–543 of your textbook.]

Identifying the Facts

1. What is the greenhouse effect? Is it natural, or is it caused by humans?

____

____

____

____

2. What role do coccolithophores play in the greenhouse effect?

____

____

____

3. How do coccolithophores cause clouds to form?

____

____

____

4. By causing clouds to form, how do coccolithophores reduce their own numbers?

____

____

____

____

5. Explain in a sentence or two how increasing the number of coccolithophores would cool off the planet.

____

____

____

____

____

Interpreting the Story

6. Why would global warming or natural changes in the planet's temperature be a problem?

____

____

____

____

____

7. In order to induce rainfall, scientists sometimes practice "cloud seeding," in which precipitation is artificially promoted. If this were done over the open ocean, what might be the immediate effect on coccolithophore populations? Explain your answer.

____

____

____

____

____

____

Thinking a Little Harder

8. Why would it be hard to predict the effects of artificially promoting coccolithophore growth?

____

____

____

____

____

MODERN EARTH SCIENCE

Unit 8: Studying Space

Eye on the Environment:
The Great Junkyard in the Sky

[Use this worksheet with pages 640–641 of your textbook.]

Identifying the Facts

1. Where did most of the space junk that is orbiting the earth come from?

____

____

____

____

2. Why is orbiting space junk a problem?

____

____

____

____

3. How do scientists now "dispose of" satellites that have outlasted their usefulness?

____

____

____

____

4. Besides the fact that more satellites have been launched into space, why has the problem of orbiting space junk grown over time?

____

____

____

____

Interpreting the Story

5. Given that orbiting space junk may pose a threat to spacecraft, do you think that we should continue to send astronauts into space? Explain your answer.

____

____

____

____

6. How do you think scientists move satellites once the satellites are in orbit?

____

____

____

____

7. Today, some rockets are equipped with automatic fuel doors. When the rockets have accomplished their missions, the doors open to release any leftover liquid fuel. What do you think happens to this fuel?

____

____

____

____

Thinking a Little Harder

8. Why would orbiting telescopes track orbiting debris better than ground-based telescopes?

____

____

____

____

____

____

Answer Key to
Eye on the Environment
Worksheets

Contents

Answers to Unit 1: Mapping Life on Earth   21

Answers to Unit 2: Volcanoes Put a Chill in the Air   23

Answers to Unit 3: What's Mined Is Yours   25

Answers to Unit 4: Losing Ground   27

Answers to Unit 5: The Mysterious Bone Bed   29

Answers to Unit 6: A Harbor Makes a Comeback   31

Answers to Unit 7: Keeping Cool With Algae   33

Answers to Unit 8: The Great Junkyard in the Sky   35

M O D E R N   E A R T H   S C I E N C E

Unit 1: Studying the Earth

Eye on the Environment: Mapping Life on Earth

[Use this worksheet with pages 62–63 of your textbook.]

Identifying the Facts

1. What was special about the animals that were discovered in Southeast Asia?

They were previously unknown to science.

2. Why were the animals in danger?

The forest that they lived in was being quickly destroyed.

3. Name at least two methods that scientists use to identify ecosystems that may be in need of protection.

Scientists use satellite photographs to identify the size and makeup of natural areas. In addition, they conduct ground surveys of the ecosystems and interview local residents.

4. Name some types of areas with a high level of biodiversity.

Answers may include tropical rain forests, coral reefs, rivers, coastal regions, and natural grasslands.

5. What is a biodiversity map?

A biodiversity map shows how many different species exist in various locations within a specific region.

Interpreting the Story

6. In your own words, define the term *biodiversity*.

Sample answer: Biodiversity refers to the number of different species that exist in a certain area.

7. Name some types of areas that might have a low level of biodiversity.

Accept all reasonable responses. Students may list areas that have been significantly disturbed by people, such as parking lots and agricultural fields, and natural areas with extreme conditions, such as polar regions.

8. How do you think satellite photographs are used to create a biodiversity map?

In general, scientists use satellite photographs to identify environments based on their different types of vegetation. The scientists also conduct ground surveys to determine the average number of species in a given area of each different type of environment. Based on the patterns of vegetation in the photographs, scientists can then estimate the level of biodiversity that exists over the entire region.

Thinking a Little Harder

9. How is creating a biodiversity map different from creating a topographic map? How is it similar?

Sample answer: Creating a biodiversity map involves measuring the biological features of an area, whereas creating a topographic map involves measuring the physical features. Creating a biodiversity map and creating a topographic map both involve defining the geographical extent of the features being measured.

10. Could you create a biodiversity map that is also topographic? Why or why not?

A map could show information about both topography and biodiversity as long as the different types of information did not interfere with each other. For instance, different colors could be used to show biodiversity and dark contour lines could be used to show topography.

M O D E R N E A R T H S C I E N C E

Unit 2: The Dynamic Earth

Eye on the Environment:
Volcanoes Put a Chill in the Air

[Use this worksheet with pages 134–135 of your textbook.]

Identifying the Facts

1. Why do some scientists think there is a link between human activities and the earth's natural greenhouse effect?

Human activities, especially the burning of fossil fuels, may increase the amount of carbon dioxide in the earth's atmosphere. Because carbon dioxide traps heat near the earth's surface, an increase in the amount of carbon dioxide in the atmosphere would increase the earth's natural greenhouse effect.

2. What chemical change happens to sulfur dioxide once it is released into the atmosphere?

It combines with water to become sulfuric acid.

3. What were the effects of Mount Pinatubo's eruption on the temperature of the earth?

By late 1992, average global surface temperatures dropped about 0.6°C and then began to slowly recover.

4. Why did the eruption of Mount Pinatubo cause the average global temperature to change?

Scientists think that once Mount Pinatubo released sulfur dioxide into the atmosphere, the sulfur dioxide changed into sulfuric acid and prevented some solar energy from reaching the earth's surface.

Interpreting the Story

5. Why are studies of volcanoes helpful to scientists who wish to understand the greenhouse effect?

Studies of volcanoes are important because volcanoes have substantial effects on the earth's climate. The gases from volcanoes remain in the atmosphere for many years, and these gases may lessen the earth's natural greenhouse effect during that time.

6. After Mount Pinatubo erupted, the effects on the earth's temperature were not observed until more than a year had passed. Explain why the changes did not occur immediately.

The global effects from Mount Pinatubo's eruption were not observed until the

volcanic gases had spread around the earth.

7. Many scientists observed the effects of Mount Pinatubo's eruption on global climate. Form a hypothesis that these scientists might have been testing with these data.

Sample hypothesis: Volcanic eruptions have no effect on average global

temperatures. (Observable changes in global temperatures following Mount

Pinatubo's eruption would help to disprove this hypothesis.)

8. Why did the scientists monitor average global surface temperatures instead of temperatures at a specific location?

The temperatures at a specific location might have reflected local climatic

factors, such as the moderating effect of a nearby body of water. By monitoring

average global temperatures, scientists could be more confident that the effects

they observed were due to the eruption.

9. What are some other natural phenomena that scientists might observe in order to study global climate changes?

Accept all reasonable responses. Answers may include large-scale weather

patterns, forest fires, meteor impacts, or changes in the sun's energy output.

Thinking a Little Harder

10. How would you determine whether the 1991 Mount Pinatubo eruption affected the average monthly temperatures in your community?

Students would need to find out the average monthly temperatures over the

course of several years for their community. They would then need to compare

those figures with the average monthly temperatures that were observed in late

1991 and 1992. If the monthly temperatures from 1991 and 1992 seem distinctly

lower than normal, then the eruption of Mount Pinatubo may have affected the

climate of their community.

M O D E R N   E A R T H   S C I E N C E

Unit 3: Composition of the Earth

Eye on the Environment: What's Mined Is Yours

[Use this worksheet with pages 214–215 of your textbook.]

Identifying the Facts

1. If the environmental costs of mining operations are so high, why do the mines exist?

Our need for the resources that come from mines is great.

2. What are some of the resources that come from mines?

The resources that come from mines include coal, copper, phosphate, uranium, precious metals, and gemstones.

3. Name at least one negative short-term effect of a mining operation.

Answers may include the observations that mining operations can destroy vegetation, divert streams, and cause flooding, droughts, or the infestation of nearby residences with snakes and other pests.

4. Name at least one negative long-term effect of a mining operation.

One negative long-term effect is the deterioration of waterways due to acid and other pollutants. The acid and pollutants also make the land near the mining site unsuitable for many plants and animals.

5. What are some organisms that are used in reclamation efforts?

Answers may include grasses, native trees, and earthworms.

6. Why do mining companies sometimes go out of business?

Mining companies may not be able to afford the reclamation procedures that are necessary once mining operations have ended. Students may also point out that mining companies might go out of business because the resource they mine becomes depleted over time.

Interpreting the Story

7. Besides mining operations, what other kinds of activities might necessitate reclamation?

Accept all reasonable responses. Answers might include landfill sites or

agricultural land.

8. Why are living organisms helpful in reclamation efforts?

To create a healthy ecosystem at a reclaimed mine, a variety of organisms must

be present to establish natural nutrient and energy cycles.

9. Why does recycling help to lessen the problems associated with mining?

Recycling makes it possible to reuse resources instead of extracting more

resources from the earth.

Thinking a Little Harder

10. In areas where old mines have never been reclaimed, the original mining companies have often long been out of business. In these cases, who do you think should have to pay the costs of reclaiming the mines?

Accept all reasonable responses.

M O D E R N   E A R T H   S C I E N C E

Unit 4: Reshaping the Crust
Eye on the Environment: Losing Ground

[Use this worksheet with pages 318–319 of your textbook.]

Identifying the Facts

1. Why are concerns about food shortages becoming more serious with each passing day?

The earth's usable topsoil is being constantly eroded away while, at the same time, the world's population is steadily growing.

2. On a small scale, how does the erosion of topsoil occur?

Bits of soil are dislodged when raindrops hit the ground. This soil may then be carried away by wind or moving water.

3. Why is soil in a forest less likely to be eroded than soil in an open, agricultural field?

Forest soil is protected by the layers of vegetation above it. Also, it is held in place by the roots of the vegetation. In a field, there is less protection from the full force of the weather.

4. Where does soil go once it has been dislodged from its original location?

Much of it is carried downwind or downstream to be deposited elsewhere. A large amount of topsoil is eventually deposited in the ocean.

5. Name four agricultural techniques that can reduce erosion in fields.

no-till farming; ridge planting; contour plowing; and crop rotation

Interpreting the Story

6. Which is less susceptible to erosion, a natural grassland or a suburban front yard? Explain your answer.

Answers may vary. In most cases, a natural grassland would have more vegetation protecting the soil than a front yard would. Thus, the front yard would be more susceptible to erosion.

7. How could a suburban front yard be changed to reduce the risk of erosion?

Answers will vary. A typical answer would be that with the planting of shrubs, trees, and other vegetation, the soil would be held in place by deeper roots, thus protecting the soil from some erosion. Also, letting the grass grow a little bit longer would offer the soil more protection from wind and rain.

8. Would erosion be greater on a level surface or a sloped surface (such as a field on the side of a hill)? Explain your answer.

Erosion would be greater on a sloped surface because gravity exerts a downward pull that causes dislodged soil and rock fragments to move downslope. On steep slopes, erosion would be even greater because steep slopes tend to have thin or poorly developed soil that cannot support dense plant growth.

Thinking a Little Harder

9. What are some factors that might keep a farmer from using sustainable farming methods?

Accept all reasonable responses. Answers might include a lack of knowledge of sustainable farming practices or the additional cost of employing these methods.

M O D E R N   E A R T H   S C I E N C E

Unit 5: The History of the Earth

Eye on the Environment: The Mysterious Bone Bed

[Use this worksheet with pages 384–385 of your textbook.]

Identifying the Facts

1. Why is the bone bed referred to as a fossil treasure chest?
   The bone bed contains fossils from a wide variety of organisms, and the fossils are well preserved. The bone bed is large and densely packed with fossils.

2. Beside mammals, what other types of organisms are represented by the Kipsaramon fossils?
   turtles, crocodiles, and trees

3. If a scientist discovered a fossil skeleton that was articulated, how would it be different from the fossils found at Kipsaramon?
   The bones of the skeleton would still be arranged in roughly the same way as they were when the animal was alive. The Kipsaramon fossils are disarticulated, which means that the bones of individual animals are jumbled and do not conform to their normal skeletal arrangement.

4. Why are the fossils shipped to Nairobi?
   In Nairobi, the fossils can be slowly extracted and cleaned without being left vulnerable to weather and people.

5. What is distinctive about fossils that have been deposited by a predator?
   The fossils show damage, such as tooth marks, or other signs of a predator's presence.

Interpreting the Story

6. Why is it helpful to find a large population of a single species in a bed bone?
   By studying many individuals of the same species, scientists can get a better idea of the overall characteristics of the species. With just one individual to study, scientists would have no way of knowing whether that individual was unusually large or small, for instance. Comparing many individuals may also reveal how members of the species grew and interacted.

7. From the list of organisms that were present at the site, what do you think the environment might have looked like?

The environment was probably mixed. The turtles and crocodiles indicate the presence of water; the antelopes and rhinos indicate some areas of grassland; and the wood, squirrels, and apes indicate the presence of trees.

8. Do you think scientists would be likely to find dinosaur fossils at this site? Explain your answer.

This site is much too recent (about 15.5 million years ago) for dinosaurs to have lived there.

Thinking a Little Harder

9. Propose a hypothesis that explains how the Kipsaramon bone bed may have formed. How would you test your hypothesis?

Accept all reasonable responses. Students' hypotheses should explain why the bones are so closely packed together and disarticulated while showing no signs of wear or damage.

M O D E R N   E A R T H   S C I E N C E

Unit 6: Oceans

Eye on the Environment:
A Harbor Makes a Comeback

[Use this worksheet with pages 450–451 of your textbook.]

Identifying the Facts

1. For how long was sewage dumped directly into Boston Harbor?

Sewage was dumped into Boston Harbor from at least the late 1700's until nearly the end of the 1900's, or over 200 years.

2. What were the effects on harbor organisms?

Over time, nearly all of the harbor organisms disappeared from the harbor.

3. In the 1970's, what facilities were in place to treat sewage from Boston? What facilities were added over the next two decades?

In the 1970's, a primary treatment plant was used to remove some of the solid materials from the city's sewage. By the end of the 1990's, primary and secondary facilities were added to remove inorganic and organic solids from the sewage and to disinfect the water before it was discharged.

4. What positive changes in Boston Harbor might a diver observe?

A diver might observe the return of fish and other organisms to the harbor. Also, a reduction in the gray substance on the harbor floor might be noticeable.

5. What is a watershed?

A watershed of a body of water consists of land from which runoff enters the body of water.

M O D E R N   E A R T H   S C I E N C E

Interpreting the Story

6. How did sewage cause outbreaks of typhoid and cholera among Boston residents?

   The sewage washed up along the shore of the city, and bacteria and other

   microorganisms in the sewage carried disease to the residents.

7. Why do you think the input of scientists was necessary when citizens and public officials wished to correct the problems in Boston Harbor?

   Scientists were able to provide information on the best ways to proceed with

   the cleanup project. Also, scientists were able to perform experiments to

   monitor the changes in the harbor once the project was under way.

8. When a federal judge ordered the state of Massachusetts to comply with the Clean Water Act, why do you think the state was given so much time? In other words, why didn't the judge order them to comply immediately?

   Such a large cleanup project would require a great deal of time and money to

   complete, not to mention the research that was necessary as background work.

   The judge probably recognized this fact when he or she set the deadline.

Thinking a Little Harder

9. Why would the health of Boston Harbor depend on the health of the surrounding watersheds?

   The watersheds that surround Boston Harbor discharge their water into the

   harbor. Therefore, any pollution that contaminates these watersheds would

   eventually flow into the harbor and contaminate the waters of the harbor.

M O D E R N   E A R T H   S C I E N C E

Unit 7: Atmospheric Forces

Eye on the Environment: Keeping Cool With Algae

[Use this worksheet with pages 542–543 of your textbook.]

Identifying the Facts

1. What is the greenhouse effect? Is it natural, or is it caused by humans?

The greenhouse effect refers to how the earth's atmosphere keeps heat trapped near the planet's surface. It is a natural process, but some scientists think that human activities may make the greenhouse effect more pronounced.

2. What role do coccolithophores play in the greenhouse effect?

Coccolithophores absorb carbon dioxide from the air. Carbon dioxide is one of the gases that causes the greenhouse effect.

3. How do coccolithophores cause clouds to form?

Coccolithophores release dimethyl sulfide (DMS) into the air, which causes water vapor to condense and form clouds.

4. By causing clouds to form, how do coccolithophores reduce their own numbers?

When clouds form, they reduce the amount of sunlight that hits the water. Because coccolithophores rely on photosynthesis for food, this reduction in sunlight causes a decrease in the number of coccolithophores.

5. Explain in a sentence or two how increasing the number of coccolithophores would cool off the planet.

Increasing the number of coccolithophores would increase the number of clouds, causing more sunlight to be reflected back into space. Increasing the numbers of coccolithophores would also decrease the amount of carbon dioxide in the atmosphere, causing a reduction in the greenhouse effect.

M O D E R N   E A R T H   S C I E N C E

Interpreting the Story

6. Why would global warming or natural changes in the planet's temperature be a problem?

Changes in the planet's temperature might cause rapid changes in natural

ecosystems. These changes might be destructive to living organisms,

including humans.

7. In order to induce rainfall, scientists sometimes practice "cloud seeding," in which precipitation is artificially promoted. If this were done over the open ocean, what might be the immediate effect on coccolithophore populations? Explain your answer.

Precipitation would lead to cloud dissipation. A reduction in the number of

clouds would allow more sunlight to reach the water, thus increasing the size

of coccolithophore populations.

Thinking a Little Harder

8. Why would it be hard to predict the effects of artificially promoting coccolithophore growth?

Coccolithophores probably interact with their environment and other organisms

in a variety of ways. While scientists may understand some of these interactions,

many others remain unknown.

M O D E R N   E A R T H   S C I E N C E

Unit 8: Studying Space

Eye on the Environment:
The Great Junkyard in the Sky

[Use this worksheet with pages 640–641 of your textbook.]

Identifying the Facts

1. Where did most of the space junk that is orbiting the earth come from?

Most of the pieces of orbiting space junk are the remains of once-useful

satellites.

2. Why is orbiting space junk a problem?

The space junk is a problem because useful spacecraft may collide with the junk

and be damaged.

3. How do scientists now "dispose of" satellites that have outlasted their usefulness?

Scientists move the satellites either farther out into unused orbits or closer in,

where they burn up as they reenter the atmosphere.

4. Besides the fact that more satellites have been launched into space, why has the problem of orbiting space junk grown over time?

Pieces of orbiting space junk have collided and have broken into smaller pieces

that in turn remain in orbit. Over time, the number of pieces in orbit has grown.

Interpreting the Story

5. Given that orbiting space junk may pose a threat to spacecraft, do you think that we should continue to send astronauts into space? Explain your answer.

Accept all reasonable responses. Some students may feel that the benefits to be

gained from studies conducted in space are worth the risks. Others may feel that

the risks are too high.

6. How do you think scientists move satellites once the satellites are in orbit?

Scientists move the satellites by remote control. The scientists might also program the satellites ahead of time.

7. Today, some rockets are equipped with automatic fuel doors. When the rockets have accomplished their missions, the doors open to release any leftover liquid fuel. What do you think happens to this fuel?

Because of the lack of pressure in space, the liquid fuel instantly vaporizes and disperses into space as a gas.

Thinking a Little Harder

8. Why would orbiting telescopes track orbiting debris better than ground-based telescopes?

Orbiting telescopes would be closer to the debris, so their images of the debris would be more detailed. Also, with ground-based telescopes, one would be viewing the debris through the atmosphere, causing possible distortions in the images.